Benthic Trawls

by Elise Hansen

ISBN-13: 978-1511968270

ISBN-10: 1511968273

Printed in the USA

The text type was set in Adobe Garamond Pro

Book design by Derek Hansen

Special thanks to my mentor and advisor
Amy Westman
an oceanographer

United States Antarctic Program
Antartic Support Contract (Lockeed Martin)
Research Vessle Ice Breaker (Nathanial B. Palmer)

as part of the
Scientists in the Classroom Program

Table of Contents

What is a benthic trawl?

A benthic trawl is an enormous net dragged across the ocean floor. The net is held *aloft** with large floats, and kept down with a heavy steel trawl door and weights. The pocket of the net is known as the "codend." The drag marks left behind from the trawl are caused by the rollers, up to five tonnes each, which bring up a dust cloud from the sea floor that can last for days. Benthic trawling has been used for over 500 years.

*See the glossary on page 22 for the definition of bold and italic words.

FUN FACT

The net opening of a trawl can get up to half an acre long! That's about as long as 405 bathtubs lined up head to tail.

CODEND

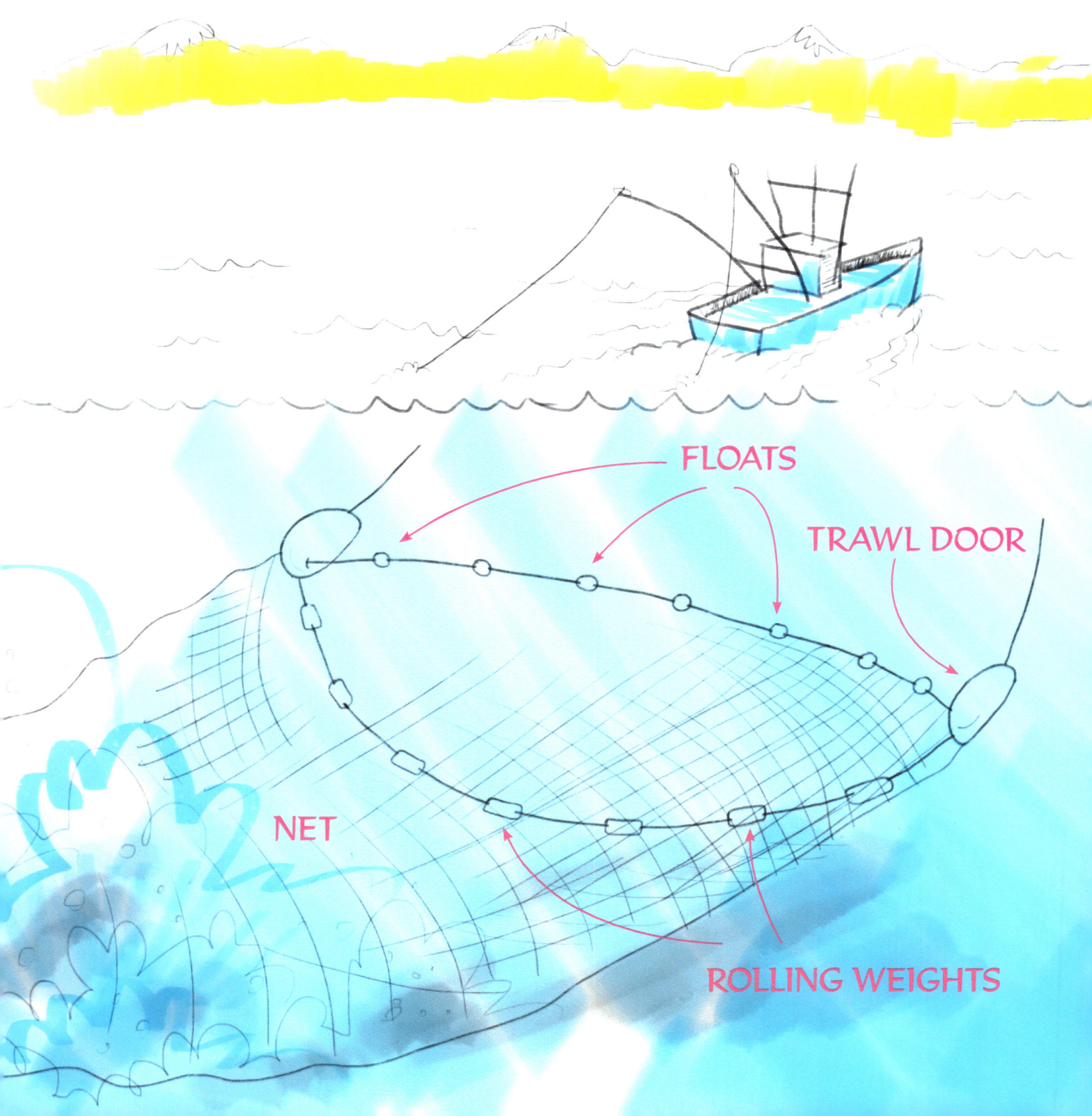

FLOATS

TRAWL DOOR

NET

ROLLING WEIGHTS

Benthic Zone

The benthic zone, in easy terms, is the ocean floor. A trawl is called a "benthic" or "bottom" trawl because the net drags across the bottom of the ocean. Because the benthic zone is so deep, sunlight does not reach that depth. Therefore, all of the plants that live on the ocean floor do not get energy from sunlight through *photosynthesis*, but from minerals that travel down from shallow waters. The fish in the benthic zone receive food from the plants.

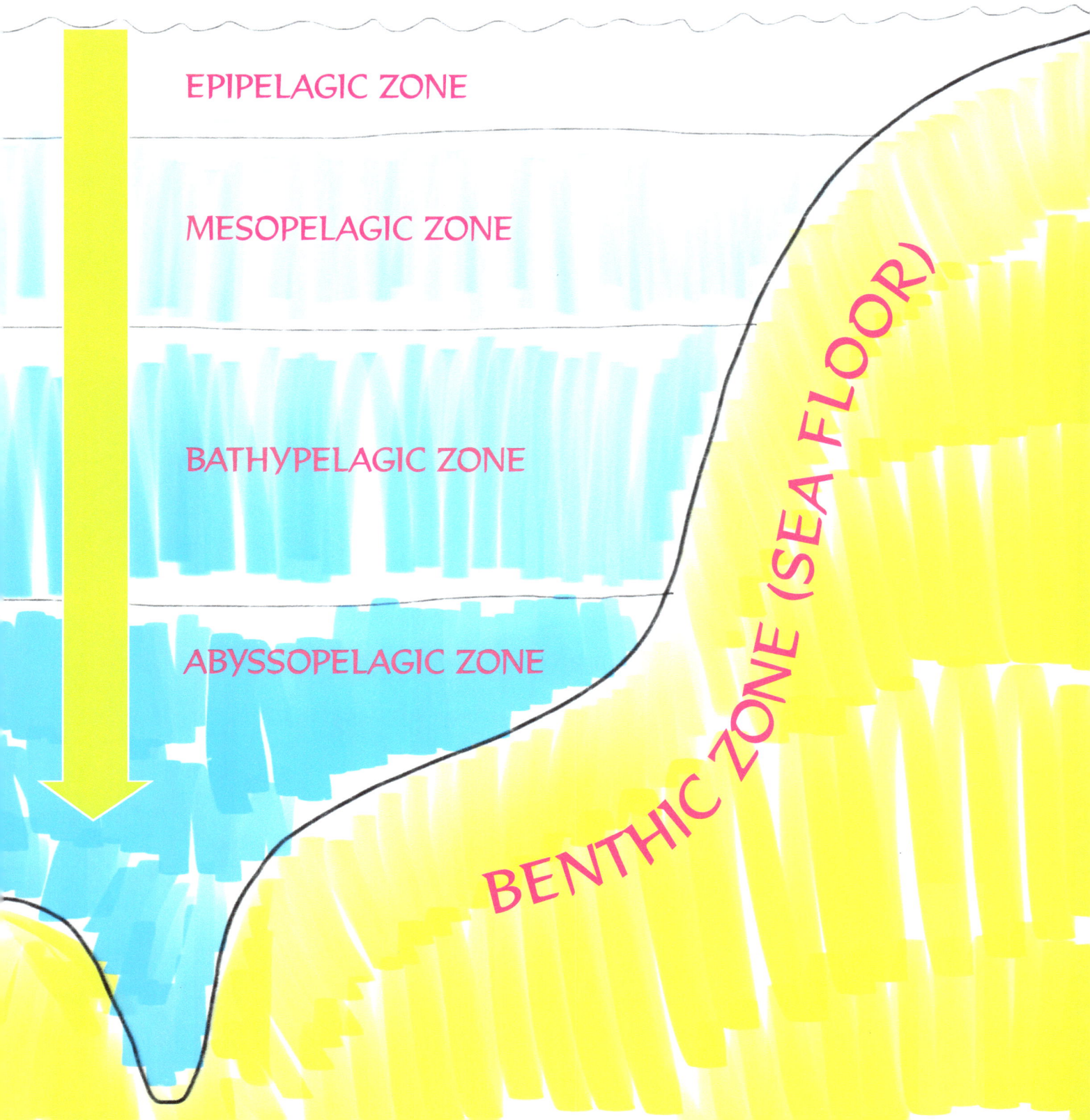

Benthos

Benthos is the name of the *organisms* that live on the benthic zone. Part of the purpose of trawling is to fish benthos for food. When a trawl rolls over an *inhabited* area of the ocean, the net comes out of the water full of all sorts of benthos, such as rock fish, cod, flounder, and shrimp. The picture to the right shows other benthos you can find deep in the ocean. Through the collection of benthos, scientists can study the animals and learn more about the ocean.

CRAB

CLAMS

Fishing

Benthic trawling is a popular and *efficient* way of fishing, because of how much *game* a trawl can catch in one run. Many countries around the world use benthic trawls as one of their main ways to catch fish. Benthic trawling is used 54% of the time between all fishing methods. The graph shown to the right demonstrates the percentage of every method. Benthic trawling is by far the largest fishing method. The second most common method is used less than half the amount of time trawling is used.

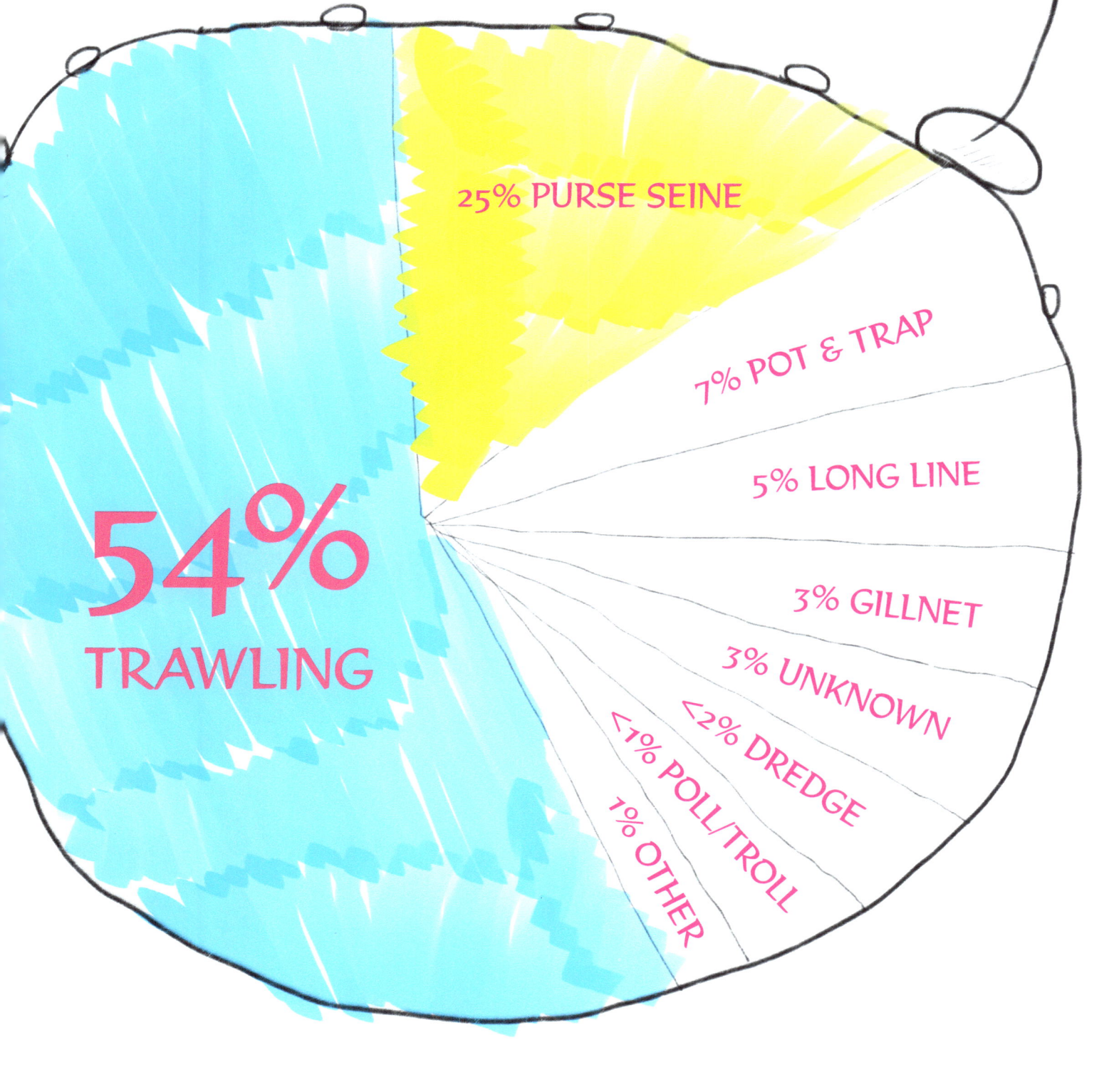

Science

Scientists from around the *globe* use the *outcome* of trawls in many different levels of science. Using the benthos of a trawl, animals and plants can be studied at deeper levels. The picture to the right shows and example of a drawing of a lobster caught in a trawl. Scientists used the lobster to learn the difference between *male* and *female* lobsters. Everything from dust, plants, trash, and rocks can be *analyzed* by *oceanographers*.

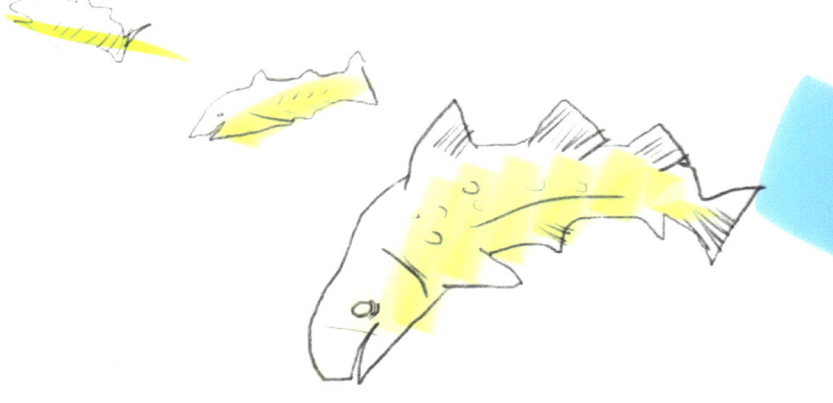

Destruction

Although trawls can be useful and *educational* at high levels, they are also destructive. The weights on the bottom of the trawl tear up *vegetation* and coral reefs that take years to grow. *Habitats* are destroyed and *marine* animals are killed. *Approximately* 8 million acres of land is destroyed per years because of benthic trawls. That's about twice the area of the United States!

Banning

Because of all the destruction trawls have caused, trawling is *gradually* becoming *outlawed* around the world. Canada, Australia, Brazil, and China are several countries that have banned areas from bottom trawling. Multiple *organizations*, such as Oceana, the Marine Conservation Institute, and the Institute of Marine Research are working at *eliminating* benthic trawls altogether. There are hopes that in the future, the benthic zone will no longer be *threatened* and life will *flourish* at the bottom of the sea. The ocean is a world of undiscovered science and nature, and we want to keep the sea alive for *generations*!

OCEANA

MARINE
CONSERVATION
INSTITUTE

INSTITUTE OF MARINE RESEARCH
HAVFORSKNINGSINSTITUTTET

Glossary

Aloft: up in the air

Analyze: to examine the structure of something

Approximately: about

Benthic: the bottom of the ocean

Benthos: animals that live at the bottom of a body of water

Educational: relating to education

Efficient: completing a great task with the smallest amount of work

Eliminate: getting rid of

Female: girl

Flourish: grow well

Game: animals hunted for food

Generation: all the people grown and living at the same time

Globe: the earth or world

Gradually: slowly, at a steady pace

Habitat: the natural home of a plant or animal

Inhabit: live in an area

Male: boy

Marine: found in or produced by the sea

Oceanographer: a scientists who studies the ocean

Organisms: Any living thing

Organization: a group of people with a particular purpose

Outcome: the way a thing turns out

Outlawed: banned or illegal

Photosynthesis: The process in which plants gather sunlight and convert the light into energy

Threatened: cause something or someone to be in danger of being hurt

Bibliography

Banerjee, Gargi. "Tourism (Pollution): Not so Incredible!" Follow Green Living. Follow Green Living, 12 May 2014. Web. 23 Dec. 2014.

"Benthic Zone." Wikipedia. Wikimedia Foundation, n.d. Web. 23 Dec. 2014.

"Bottom Trawling." Bottom Trawling. N.p., n.d. Web. 22 Dec. 2014.

"Bottom Trawling Impacts On Ocean, Clearly Visible From Space." ScienceDaily. ScienceDaily, n.d. Web. 23 Dec. 2014.

"Destructive Fishing." Marine Conservation Institute. Marine Conservation Institute, n.d. Web. 22 Dec. 2014.

"The Fishing Industry and Our Oceans." Middle School Science at Synergy School. Middle School Science at Synergy School, 16 May 2012. Web. 23 Dec. 2014.

"Gulf Coast Preservation Society- Bottom Trawlers." Gulf Coast Preservation Society- Bottom Trawlers. Gulf Coast Preservation Society, n.d. Web. 23 Dec. 2014.

"The Impact of Trawling." Institute of Marine Research. Institute of Marine Research, n.d. Web. 23 Dec. 2014.

—

"Infographic: What Is Trawling?" Oceana. Oceana, n.d. Web. 22 Dec. 2014.

"Manual of the International Bottom Trawling Surveys." Series of ICES Survey Protocols (n.d.): 1-68. Ices.dk. Web.

"Pelagic Environment Animals." Images & Pictures. N.p., n.d. Web. 23 Dec. 2014.

www.ingramcontent.com/pod-product-compliance
Lightning Source LLC
Chambersburg PA
CBHW050432180526
45159CB00006B/2502